THE BOOK OF MORMON ANSWERS THE FERMI PARADOX

ILYAN KEI LAVANWAY

Ilyan Kei Lavanway
Lehigh, Kansas

THE BOOK OF MORMON ANSWERS THE FERMI PARADOX
Copyright © 2020 Ilyan Kei Lavanway

Cover image courtesy of:
https://www.wallpapertip.com/wmimgs/40-409856_earth-from-space-uhd-wallpaper-1-space-wallpapers.jpg

This written work may be freely copied, printed, shared and distributed for the enjoyment and edification of everyone.

This is not an official publication of any organization, nor does it represent any official position. This work is the sole responsibility of the author and includes the author's personal insights.

ISBN: 9798586510273

Published by Ilyan Kei Lavanway
Lehigh, Kansas, USA

The Book of Mormon Answers The Fermi Paradox

Dedicated to Doctor Matthew Perry, DMD

Thank you for reading my books and fixing my teeth and giving me a priesthood blessing. May God bless you and your family.

The Book of Mormon Answers The Fermi Paradox

This page intentionally left blank.

CONTENTS

What Is The Fermi Paradox? 1

One Side of The Fermi Paradox:
They Should Be Out There 3

Other Side of The Fermi Paradox:
Why Have We Not Discovered Them? 13

Written Works by Ilyan Kei Lavanway 19

Author Contact 23

This page intentionally left blank.

WHAT IS THE FERMI PARADOX?

Stated simply, the Fermi Paradox is a combination of the Drake Equation and the Kardashev Scale accompanied by Enrico Fermi's famous question asked in 1950: Where is everybody?

For readers not yet familiar with these terms, the Fermi Paradox acknowledges two opposing facts: Our galaxy should be teeming with intelligent civilizations. We have not discovered any empirical evidence of their existence.

Or have we? For the reader interested in considering that indeed we do know about intelligent life beyond Earth, and conspiracies trying to cover up such knowledge, I suggest the following book:

Bones In The Sand: The Extraterrestrial Connection to Earth by V. G. La Van Way, ISBN 9781547283101

Before putting my two cents worth into the topic of the Fermi Paradox, here are three links to articles explaining the Fermi Paradox in more detail. Bear in mind, these are secular perspectives. They do not in any way acknowledge the existence of God, or even suggest the possibility of Intelligent Design or Creationism as an alternative to Evolution.

https://www.space.com/25325-fermi-paradox.html

https://www.seti.org/seti-institute/project/fermi-paradox

https://www.britannica.com/story/the-fermi-paradox-where-are-all-the-aliens

ONE SIDE OF THE FERMI PARADOX: THEY SHOULD BE OUT THERE

Well, they are. While there are countless species of intelligent life indigenous to other worlds beyond Earth, I am most specifically interested in, and focused on, the existence of human beings indigenous to worlds other than Earth.

While I cannot offer empirical evidence of the existence of intelligent life beyond Earth, I assure you with every fiber of my being that there are countless worlds similar to this earth that are populated with human beings like us, working through their own mortal probation, having the ordinances of salvation and exaltation made available to them by and through Jesus Christ. He is as much their savior as he is ours. They are begotten spirit sons and daughters of the same Heavenly Father who begat our spirits.

This pattern is eternal. It does not begin or end with our Heavenly Father. However, the extension of this pattern beyond our own Heavenly Father is a discussion outside the scope of this book. For the reader interested in exploring beyond the scope of this book, please avail yourself of the following works:

Onions of Eternity by Ilyan Kei Lavanway
ISBN 9781520402222

Intelligent Universe by Ilyan Kei Lavanway
ISBN 9781494905910

Nature of The Godhead by Ilyan Kei Lavanway
ISBN 9780463012000

Circumscription Hypothesis by Ilyan Kei Lavanway
ISBN 9781505429329

Thought Log 2015.08.17.1900 by Ilyan Kei Lavanway
ISBN 9781523422173

The Grandeur of Christmas and The Son of God
by Ilyan Kei Lavanway, ISBN 9781520260082

Eternal Family Structures Among Exalted Couples
by Ilyan Kei Lavanway, not yet published
Copies available available upon request.
Contact: ilyanlavanway@gmail.com

I'm going to cite scripture. This is faith-based evidence.

Perhaps the reader may ask, why should I care? Why is any of this important? Shouldn't I just stick to Gospel basics?

Good questions. Here's a better question: Considering all the scriptures we could have, but don't have, why were the following verses included in the scriptures we do have?

Given the effort and sacrifice required of those who made these scriptures available to us, why was it so important to include these particular verses and other similar verses?

Here's another question: If we should just stick to Gospel basics, then don't you think God would just give us basic scriptures? I guess that means this is all pretty basic stuff. So, maybe we should learn to understand it.

What's the risk? I might accidentally deepen my personal relationship with my Heavenly Father, and with my Savior, Jesus Christ, and with the Holy Ghost. I might accidentally become a more converted disciple of Christ. I might even become happy! That's a risk I'm willing to take.

Speaking of Jesus Christ, the prophet Joseph Smith gave one of the most authoritative and unequivocal declarations in history regarding the existence of a divine Creator and the existence of people indigenous to other planets, and their relationship to the Creator and to us, and the fact that the creation of worlds populated with human beings has been going on long before Earth, and will continue forever:

"And now, after the many testimonies which have been given of him, this is the testimony, last of all, which we give of him: That he lives! For we saw him, even on the right hand of God; and we heard the voice bearing record that he is the Only Begotten of the Father—That by him, and through him, and of him, the worlds are and were created, and the inhabitants thereof are begotten sons and daughters unto God." (Doctrine and Covenants 76:22-24)

That one verse alone should be enough to convince any sound mind of the existence of people indigenous to other worlds—extraterrestrial intelligent life. As with all matters pertinent to man, there are multitudes of confirming revelations and additional witnesses further illuminating the facts:

"And when ye shall receive these things, I would exhort you that ye would ask God, the Eternal Father, in the name of Christ, if these things are not true; and if ye shall ask with a sincere heart, with real intent, having faith in Christ, he will manifest the truth of it unto you, by the power of the Holy Ghost. And by the power of the Holy Ghost ye may know the truth of all things." (Moroni 10:4-5)

"Have we not all one father? hath not one God created us? why do we deal treacherously every man against his brother, by profaning the covenant of our fathers?" (Malachi 2:10)

"And God spake unto Moses, saying: Behold, I am the Lord God Almighty, and Endless is my name; for I am without beginning of days or end of years; and is not this endless? And, behold, thou art my son; wherefore look, and I will show thee the workmanship of mine hands; but not all, for my works are without end, and also my words, for they never cease. Wherefore, no man can behold all my works, except he behold all my glory; and no man can behold all my glory, and afterwards remain in the flesh on the earth." (Moses 1:3-5)

"And worlds without number have I created; and I also created them for mine own purpose; and by the Son I created them, which is mine Only Begotten." (Moses 1:33)

"And the Lord God spake unto Moses, saying: The heavens, they are many, and they cannot be numbered unto man; but they are numbered unto me, for they are mine. And as one earth shall pass away, and the heavens thereof even so shall another come; and there is no end to my works, neither to my words. For behold, this is my work and my glory—to bring to pass the immortality and eternal life of man." (Moses 1:37-39)

Notice how God states that he has a reason for continuing to create worlds without number. He creates them for his own purpose. And then, he explains what his most important purpose is: to bring to pass the immortality and eternal life of man.

To put it another way, the most important purpose behind the creation of countless worlds is to afford man—and man is the literal offspring of God (see Acts 17:28-29)—a means and a place to experience a mortal probation whereby man can learn by personal experience to discern between good and evil, to exercise individual agency and stewardship, to eventually qualify for exaltation, which is godhood, to become gods, equal to our Heavenly Father (see Abraham 3:24-25; Alma 34:32-33; Doctrine and Covenants 78:5-7; 88:107; 104:17; 132:19-20; 130:22).

"Behold, the Lord hath created the earth that it should be inhabited; and he hath created his children that they should possess it." (1 Nephi 17:36)

"And he beheld many lands; and each land was called earth, and there were inhabitants on the face thereof." (Moses 1:29)

"And were it possible that man could number the particles of the earth, yea, millions of earths like this, it would not be a beginning to the number of thy creations; and thy curtains are stretched out still; and yet thou art there, and thy bosom is there; and also thou art just; thou art merciful and kind forever;" (Moses 7:30)

"Wherefore, I can stretch forth mine hands and hold all the creations which I have made; and mine eye can pierce them also, and among all the workmanship of mine hands there has not been so great wickedness as among thy brethren." (Moses 7:36)

We will come back to this passage in Moses 7:36 later. It effectively answers the Fermi Paradox all by itself, but there is more to say about it to illustrate the point clearly. Now you have a hint, if not a total giveaway.

"And I saw the stars, that they were very great, and that one of them was nearest unto the throne of God; and there were many great ones which were near unto it; And the Lord said unto me: These are the governing ones; and the name of the great one is Kolob, because it is near unto me, for I am the Lord thy God: I have set this one to govern all those which belong to the same order as that upon which thou standest." (Abraham 3:2-3)

"And where these two facts exist, there shall be another fact above them, that is, there shall be another planet whose reckoning of time shall be longer still; And thus there shall be the reckoning of the time of one planet above another, until thou come nigh unto Kolob, which Kolob is after the reckoning of the Lord's time; which Kolob is set nigh unto the throne of God, to govern all those planets which belong to the same order as that upon which thou standest." (Abraham 3:8-9)

Wait. What's all this about a Kolob being set to govern all those worlds which belong to the same order as the one upon which we stand? You mean, there really are more earths like ours? Is that what that just said?

Yes. There's a whole order — a multitude — of earths like ours with people like us being born into mortal life upon them. And if there is an order of worlds to which our earth belongs, then there must be other orders of different types of worlds not like earth. Each order must have some type of governing body (see Facsimile 2 Explanation, The Book of Abraham).

"And it is given unto thee to know the set time of all the stars that are set to give light, until thou come near unto the throne of God. Thus I, Abraham, talked with the Lord, face to face, as one man talketh with another; and he told me of the works which his hands had made; And he said unto me: My son, my son (and his hand was stretched out), behold I will show you all these. And he put his hand upon mine eyes, and I saw those things which his hands had made, which were many; and they multiplied before mine eyes, and I could not see the end thereof." (Abraham 3:10-12)

"And there are many kingdoms; for there is no space in the which there is no kingdom; and there is no kingdom in which there is no space, either a greater or a lesser kingdom. And unto every kingdom is given a law; and unto every law there are certain bounds also and conditions." (Doctrine and Covenants 88:37-38)

"Unto what shall I liken these kingdoms, that ye may understand?" (Doctrine and Covenants 88:46)

"Behold, I will liken these kingdoms unto a man having a field, and he sent forth his servants into the field to dig in the field. And he said unto the first: Go ye and labor in the field, and in the first hour I will come unto you, and ye shall behold the joy of my countenance. And he said unto the second: Go ye also into the field, and in the second hour I will visit you with the joy of my countenance. And also unto the third, saying: I will visit you; And unto the fourth, and so on unto the twelfth. And the lord of the field went unto the first in the first hour, and tarried with him all that hour, and he was made glad with the light of the countenance of his lord. And then he withdrew from the first that he might visit the second also, and the third, and the fourth, and so on unto the twelfth. And thus they all received the light of the countenance of their lord, every man in his hour, and in his time, and in his season—Beginning at the first, and so on unto the last, and from the last unto the first, and from the first unto the last; Every man in his own order, until his hour was finished, even according as his lord had commanded him, that his lord might be glorified in him, and he in his lord, that they all might be glorified. Therefore, unto this parable I will liken all these kingdoms, and the inhabitants thereof—every kingdom in its hour, and in its time, and in its season, even according to the decree which God hath made." (Doctrine and Covenants 88:51-61)

This page intentionally left blank.

OTHER SIDE OF THE FERMI PARADOX: WHY HAVE WE NOT DISCOVERED THEM?

This is where the Book of Mormon brings to bear with perfect clarity the bottom line as to why we have no empirical, widely acknowledged evidence of inhabitants, particularly human inhabitants, indigenous to worlds other than our own earth. This takes some simple deductive reasoning to extract. Remember Moses 7:36? This is where one of the meanings of that verse comes into play.

While reading the following verses, you will undoubtedly comprehend that Jesus is speaking to the Nephite nation in Ancient America, and he is explaining to the Nephites why the Jews don't know anything about them, and why the Jews don't know anything about the Ten Tribes, and why the Jews don't know anything about other sheep—other sheep meaning other people whom Christ shepherds.

Don't merely gloss over these verses like you've seen them so many times before. As you read them now, imagine Jesus visiting a distant world and speaking to a vast conference attended by the inhabitants of other earths. Imagine Jesus referring to our earth in the same manner he referred to the Jews in his address to the Nephites.

"And it came to pass that when Jesus had made an end of praying he came again to the disciples, and said unto them: So great faith have I never seen among all the Jews; wherefore I could not show unto them so great miracles, because of their unbelief. Verily I say unto you, there are none of them that have seen so great things as ye have seen; neither have they heard so great things as ye have heard." (3 Nephi 19:35-36)

"And behold, this is the land of your inheritance; and the Father hath given it unto you. And not at any time hath the Father given me commandment that I should tell it unto your brethren at Jerusalem" (3 Nephi 15:13-14).

"And now, because of stiffneckedness and unbelief they understood not my word; therefore I was commanded to say no more of the Father concerning this thing unto them. But, verily, I say unto you that the Father hath commanded me, and I tell it unto you, that ye were separated from among them because of their iniquity; therefore it is because of their iniquity that they know not of you. And verily, I say unto you again that the other tribes hath the Father separated from them; and it is because of their iniquity that they know not of them." (3 Nephi 15:18-20)

"Wherefore, as I said unto you, it must needs be expedient that Christ—for in the last night the angel spake unto me that this should be his name—should come among the Jews, among those who are the more wicked part of the world; and they shall crucify him—for thus it behooveth our God, and there is none other nation on earth that would crucify their God. For should the mighty miracles be wrought among other nations they would repent, and know that he be their God. But because of priestcrafts and iniquities, they at Jerusalem will stiffen their necks against him, that he be crucified." (2 Nephi 10:3-5)

Get the picture? Not yet? Okay, let's try this another way. Let's do the same thought exercise and pretend Jesus is talking to a multi-world conference being held on a distant planet. No one from our earth is attending. Jesus mentions our earth to the inhabitants of these other worlds. This time, let's imagine exactly what Jesus is saying to the people on other worlds as he talks to them about us, and explains to them why we don't know anything about them:

And it came to pass that Jesus said unto them: So great faith have I never seen among all the inhabitants of that earth which is my footstool, whereupon I performed my atonement of infinite scope and eternal efficacy, which atonement applies to all the worlds I have created and will create, and all the spirit offspring which my Father, which is your Father, has begotten and will beget. Wherefore I could not show unto the inhabitants of that earth which is my footstool so great miracles, because of their unbelief. Verily I say unto you, there are none of them that have seen so great things as ye have seen; neither have they heard so great things as ye have heard." (compare 3 Nephi 19:35-36)

And behold, these worlds are the lands of your inheritance; and the Father hath given them unto you. And not at any time hath the Father given me commandment that I should tell it unto your brethren upon that earth which is my footstool, which is the seat of my infinite and eternal atonement, which atonement availeth all the Father's children on all the worlds I have created and will create, forever more." (compare 3 Nephi 15:13-14)

And now, because of stiffneckedness and unbelief, they who dwell upon that earth which is my footstool, which is the seat of my atonement, understood not my word; therefore I was commanded of the Father to say nothing concerning you unto them. But, verily, I say unto you that the Father hath commanded me, and I tell it unto you, that ye are hidden from them because of their iniquity; therefore it is because of their iniquity that they know not of you. And verily, I say unto you again that the inhabitants of all other worlds hath the Father hidden from them; and it is because of their iniquity that they know not of them." (compare 3 Nephi 15:18-20)

Wherefore, as I said unto you, it must needs be expedient that Christ should come among the inhabitants of that earth which is his footstool, among those who are the more wicked part of that world; and they shall crucify him—for thus it behooveth our God, and there is none other world among all the worlds created by Christ that would crucify their God. For should the mighty miracles be wrought among the inhabitants of other worlds, they would repent, and know that he be their God. But because of priestcrafts and iniquities, they who dwell upon that earth which is Christ's footstool will stiffen their necks against him, that he be crucified." (compare 2 Nephi 10:3-5)

This page intentionally left blank.

WRITTEN WORKS BY ILYAN KEI LAVANWAY

Published:

The Book of Mormon Answers The Fermi Paradox (2020)
Cleansing of The Church (2020)
Dreams of the Last Days: Dragons and Clocks (2020)
Dreams of the Last Days: Mobile Tactical Temples (2017)
Dreams of the Last Days: Psychokinetic Flight and Pyrokinesis (2016)
Dreams of the Last Days: Marked and Messaged (2015)
Dreams of the last Days: Machines and Machinations (2015)
Dreams of the Last Days: Where are all the Stars? (2015)
Vanishing Room (2019)
An Aviator At Heart (2014, 2019)
Nature of the Godhead (2019)
Guidebook for Your Journey Home (2017)
Onions of Eternity (2017)
The Grandeur of Christmas and The Son of God (2016)
Paradise and Spirit Prison (2016)
What Happens at the Second Coming of Jesus Christ (2016)
Thought Log 2015.08.17.1900 (2016)

The Lowliest of Callings (2016)

Families are Meant to Last Forever (2016)

Dream Log 2015.09.18.0900 (2015)

Thought Log 2015.08.09.2100 (2015)

Baptism by Immersion in Water and in the Spirit of God (2015)

Likening My Life to the Apostle Dieter Friedrich Uchtdorf (2015)

If Christ had not Suffered (2015)

Honoring His Namesake (2015)

FEMA Camp Parent Trap (2015)

Marriage (2015)

Relying on the Lord in the Last Days (2015)

Circumscription Hypothesis (2015)

Intelligent Universe (2014)

Sevenfold (2013)

The Modern Day Gadianton Golden Boy (2012)

Post Omerican Easter (2012)

Duck Boy on the Platypus Farm (2012)

Platypus Boy on the Duck Farm (2012)

Into the Picture (2012)

Earth Sink (2010)

Chicken Poop From the Hole (2005)

Unpublished (Note: this list is not complete):

Eternal Family Structures Among Exalted Couples (2020)

Personal Impressions Scripture Study Log: 2019 Edition (2020)

100-Word Flash Fiction Collection (2018)

You Cannot Wield Your Way with a Sword of Sand (2018)

Fast Offerings: The Lord's Way of Blessing Both Giver and Receiver (2017)

Knit Together as One (2017)

Object-Generated Wormholes as Wake Effect Causing Superluminal Transit (2017)

Cinderella Cruella (2017)

Keeping the Sabbath Day Holy: Sacrament Meeting (2017)

Facebook Conversation about Adam and Eve (2016)

Dream Log 2016.12.24 0600 (2016)

The Parable of the Ten Virgins for Our Day (2016)

A Great American Dichotomy (2016)

States of the Flesh (2016)

Home Teaching: The Meat of the Gospel (2016)

Missionary Successes Visible and Invisible (2016)

Obey the Commandments of God (2016)

A Lesson from Job for Our Day (2016)

Why the Word of Wisdom Warns us to Avoid Coffee and Tea (2016)

The Atonement Affords Access to Parallel Realities (2016)

Is a New Age Deception Poisoning Some of the Elect? (2016)

Thoughts on the Benefits of Asking for a Priesthood Blessing (2016)

Soul Reacher (2012)

This page intentionally left blank.

AUTHOR CONTACT

Email:
ilyanlavanway@gmail.com

Amazon Author Page:
https://www.amazon.com/author/ilyan
https://www.amazon.com/Ilyan-Kei-Lavanway/e/B004YL1HG2

Smashwords Author Page:
https://www.smashwords.com/profile/view/ilyan

Goodreads Author Page:
https://www.goodreads.com/author/show/4751470.Ilyan_Kei_Lavanway

LinkedIn:
https://www.linkedin.com/in/ilyanlavanway/

Twitter:
@ilyanlavanway
https://twitter.com/ilyanlavanway
@dare2unveil
https://twitter.com/dare2unveil

Website:
https://www.cajunostomy.com

This page intentionally left blank.

www.ingramcontent.com/pod-product-compliance
Lightning Source LLC
Chambersburg PA
CBHW031600210526
45464CB00003B/1362